OIL SPILL!

STAGE 2

BY MELVIN BERGER
ILLUSTRATED BY PAUL MIROCHA

HarperCollins*Publishers*

The art for this book was done in opaque watercolor and colored pencil on Arches Hot Press paper.

The *Let's-Read-and-Find-Out Science* book series was originated by Dr. Franklyn M. Branley, Astronomer Emeritus and former Chairman of the American Museum–Hayden Planetarium, and was formerly co-edited by him and Dr. Roma Gans, Professor Emeritus of Childhood Education, Teachers College, Columbia University. Text and illustrations for each book in the series are checked for accuracy by an expert in the relevant field. For a complete catalog of Let's-Read-and-Find-Out Science books, write to HarperCollins Children's Books, 10 East 53rd Street, New York, NY 10022.

Let's Read-and-Find-Out Science is a registered trademark of HarperCollins Publishers.

Library of Congress Cataloging-in-Publication Data
Berger, Melvin.
    Oil spill! / by Melvin Berger ; illustrated by Paul Mirocha.
        p.   cm. — (Let's-read-and-find-out science. Stage 2)
    Summary: Explains why oil spills occur and how they are cleaned up and suggests strategies for preventing them in the future.
    ISBN 0-06-022909-8. — ISBN 0-06-022912-8 (lib. bdg.)
    ISBN 0-06-445121-6 (pbk.)
    1. Oil spills—Environmental aspects—Juvenile literature.   2. Oil pollution of the sea—Juvenile literature.   [1. Oil spills.   2. Pollution.]
    I. Mirocha, Paul, ill.   II. Title.   III. Series.
    TD427.P4B46   1994   92-34779
    363.73'82—dc20   CIP
                                                                                            AC

1   2   3   4   5   6   7   8   9   10   ❖
First Edition

OIL SPILL!

THE NIGHT OF MARCH 24, 1989, was dark and cold. A huge, black oil tanker glided out of the port of Valdez, Alaska. Painted on its bow was its name—*Exxon Valdez*.

The *Exxon Valdez* floated low in the water. About fifty million gallons of crude oil weighed it down.

The huge tanker slowly sailed out into Prince William Sound. And then, suddenly, it happened.

CRUNCH!

The ship slammed into an underwater reef!

Thick, black oil flowed out of the *Exxon Valdez*'s smashed tanks. It poured into the dark water.

The *Exxon Valdez* spilled out 11 million gallons of oil. That much oil could fill over 1,000 big swimming pools.

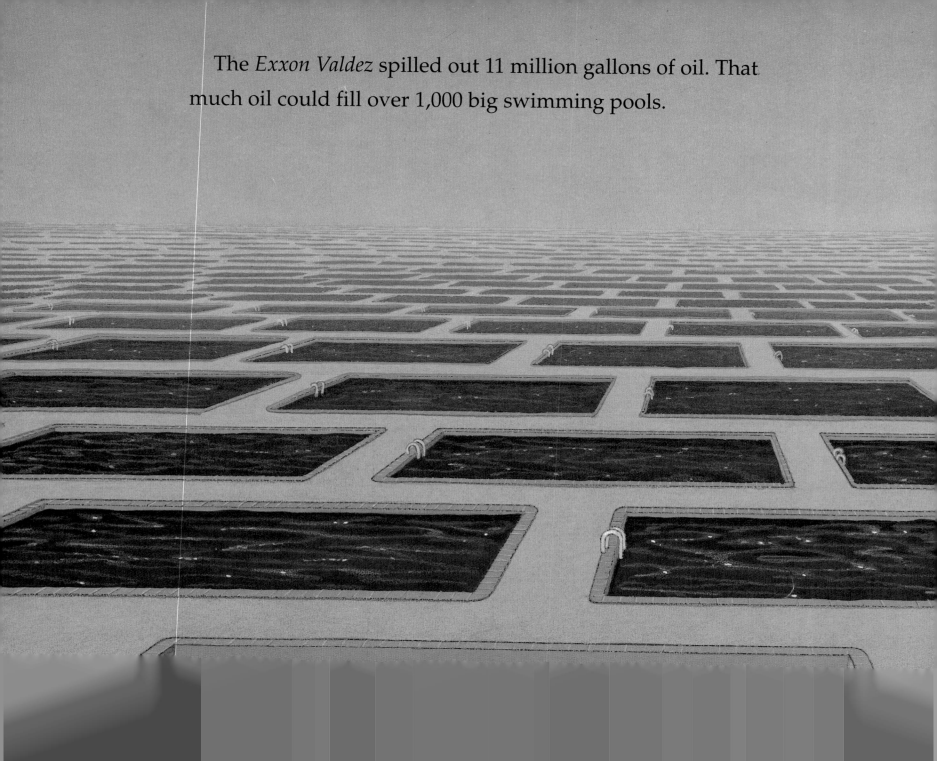

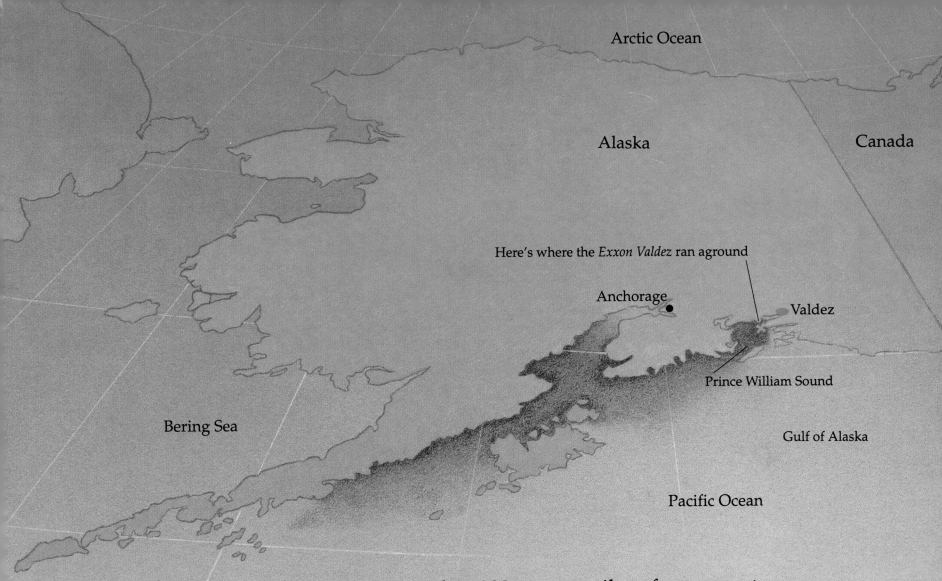

Arctic Ocean

Alaska

Canada

Here's where the *Exxon Valdez* ran aground

Anchorage

Valdez

Prince William Sound

Bering Sea

Gulf of Alaska

Pacific Ocean

The sticky oil soon covered 11,000 square miles of ocean water. That is an area as big as the state of Maryland.

It damaged about 1,250 miles of Alaska's coastline. That is longer than the entire Atlantic coast of the United States.

The oil stuck to the feathers of many ducks, geese, and other seabirds. The birds couldn't swim or fly. Over 300,000 died.

People of all ages volunteered to help in the clean-up of the spill.

Oil got into the bodies of fish, shrimp, and crabs. No one knows how many of them were killed.

salmon fry

Sea otters, sea lions, harbor seals, and killer whales swallowed
oil. They breathed the poisonous fumes. Their bodies were coated
with oil. Thousands of these marine mammals died.

The spill from the *Exxon Valdez* was one of the worst in our country's history. But it was not the only one. An oil spill occurs somewhere in the world almost every day of the year.

Oil spills have many causes. Some are accidents.

- A tanker like the *Exxon Valdez* runs aground or collides with another ship.
- Workers make mistakes as a tanker is being loaded or unloaded.
- An undersea oil well starts to leak.
- A tank or a pipe breaks at a shore oil terminal.

Some oil spills occur on purpose. The tanker captain tells the workers to clean out the tanks. Sometimes they flush the old oil into the sea.

Tanker collides with another tanker

Undersea oil well starts to leak

Pipe breaks at shore oil terminal

The causes of oil spills differ. But the result is the same. The oil spreads out. It floats on top of the water.

Experts on oil spills rush to the scene. They start to clean up the mess.

Their first job is to stop the oil from spreading. They put a boom around the spill. The boom is like a collar. It keeps the oil in one place.

For small spills, the experts may call for a skimmer. There are several kinds of skimmers. One type works like a giant vacuum cleaner. It sucks up the oil from the water. Sometimes the oil the skimmer collects can be used again.

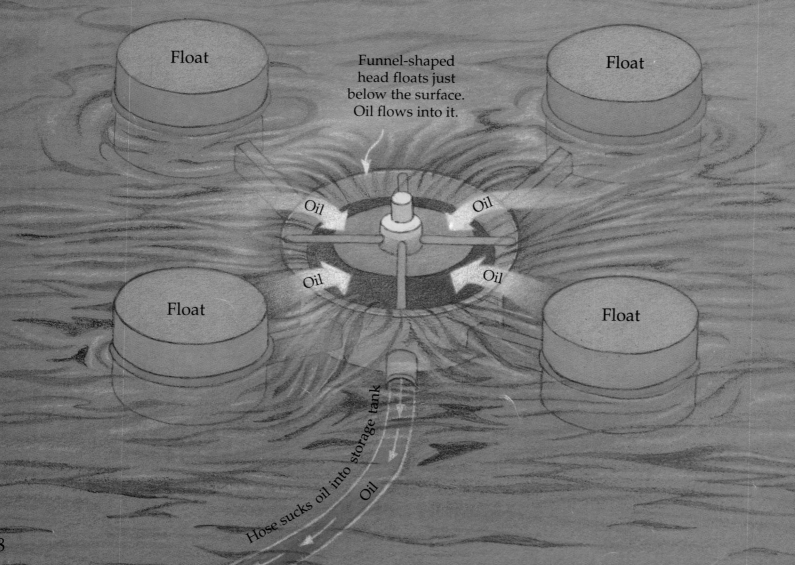

Float

Float

Funnel-shaped head floats just below the surface. Oil flows into it.

Oil

Oil

Oil

Oil

Float

Float

Hose sucks oil into storage tank

Oil

Bags filled with soaked-up oil

Pad

Oil

Water

For some small spills, experts place special pads on top of the oil. The pads are like sponges. They soak up the oil. Then they have to get rid of the soaked-up oil.

19

For larger spills, the experts may set
the oil on fire. But the fire sends smoke
and gas into the air and leaves ash in

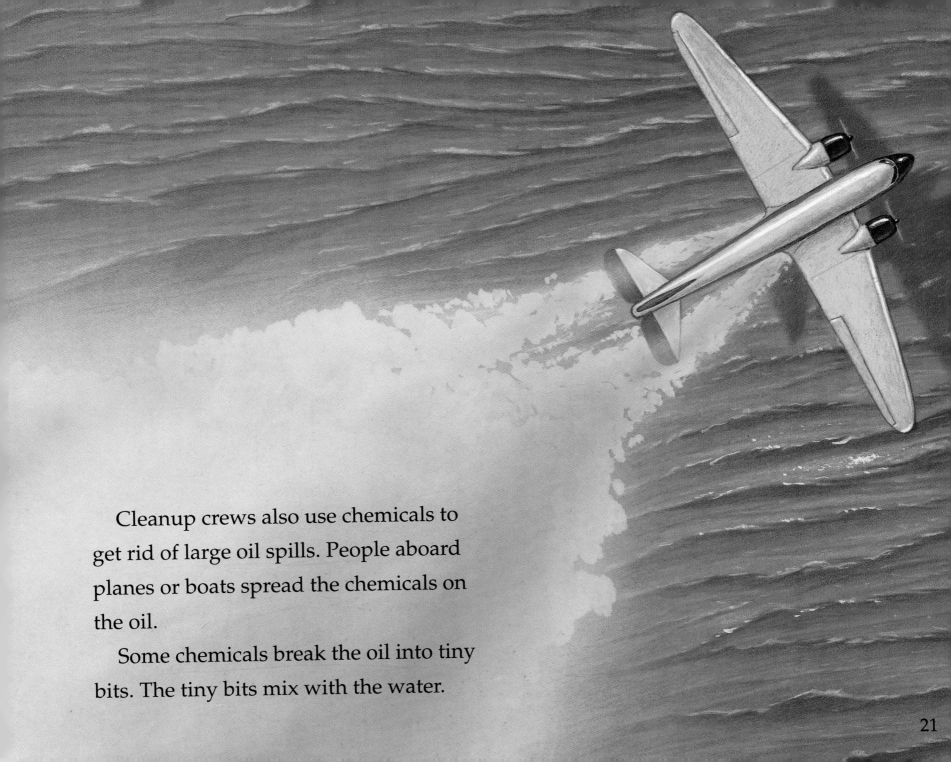

Cleanup crews also use chemicals to get rid of large oil spills. People aboard planes or boats spread the chemicals on the oil.

Some chemicals break the oil into tiny bits. The tiny bits mix with the water.

21

Other chemicals make the oil come together. The oil forms a layer like a sheet of rubber. One type of skimmer lifts the "sheet" of oil and oily debris out of the water.

Chemicals make the oil less harmful. But they add poisons to the water.

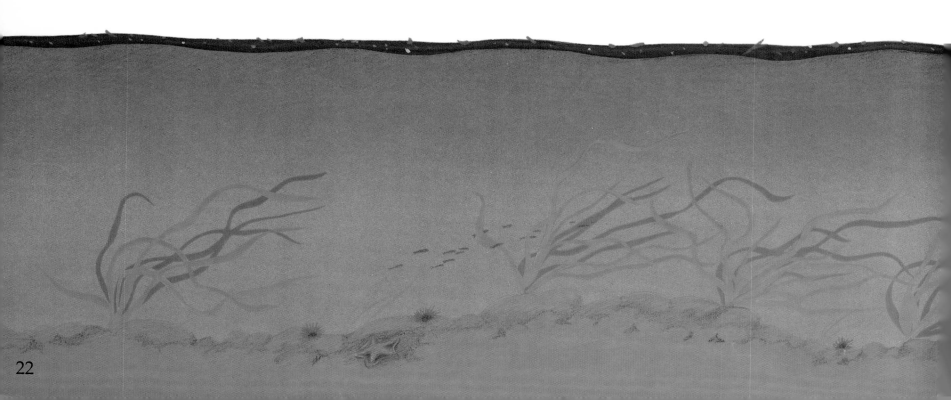

Oil

Debris

Motor

Oil

Pump

Water

In time, the oil from most spills drifts up onto the shore. Scientists spray the rocks and sand with hot water to wash the oil back into the sea. But the hot spray may also push the oil farther into the rocks and sand. Here the oil can harm plants and animals that live on the shore.

Scientists sometimes add bacteria to the oil along the shore. The bacteria "eat" the oil and change it into harmless substances. But it would take huge amounts of bacteria to get rid of a big oil spill.

Sometimes the experts decide that no action is the best way to treat an oil spill. The wind and waves mix the oil and water together. It is like mixing oil and vinegar to make salad dressing. In time, much of the oil disappears.

Oil spills are major disasters. Slowly scientists are learning how to clean them up. They are learning how to prevent spills. How can we help to prevent oil spills, too?

How to Help Prevent Oil Spills

We can use less oil. If we use less oil, there will be fewer oil tankers in the oceans. Then the chance of oil spills will not be as great.

One way to save oil is to use less electricity. Electricity is often produced by burning oil. Less electricity means less oil.

Another way to cut oil needs is to use less gasoline, which is made from oil. That means driving smaller cars and staying within the speed limit.

We also can write letters to members of Congress. Tell them we want new laws to prevent oil spills:

- Oil tankers should have double hulls. If the outer hull is damaged, the inner one will still hold the oil.

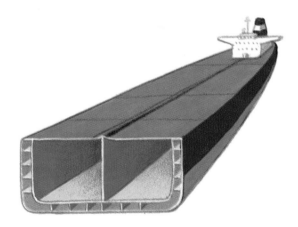

- Tankers should have the most modern radar, radio, and other safety systems. This will help to prevent collisions.
- Experts should teach tanker crews how to handle oil spills.
- Booms, skimmers, chemicals, and bacteria should be kept ready for emergencies all around the world.
- Oil companies should be helped to find new oil fields in the United States. They should not just depend on oil shipped from abroad.

Together we can make our seas and shorelines clean
and beautiful again!

Senator _____
U.S. Senate
Washington, DC 20510

Dear Senator,
I am worried about oil spills.
They kill birds, fish, and other animals that
live in the sea. We must protect our
Earth. We need new laws to help prevent
oil spills.
        Sincerely,
            Anna

476204

$14.89

DATE			